Measures

Angle 1

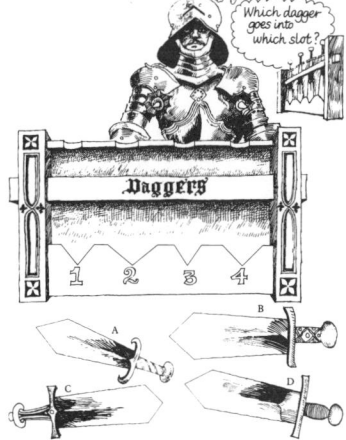

Dagger A goes into slot **3**.
Dagger B goes into slot **4**.
Dagger C goes into slot **2**.
Dagger D goes into slot **1**.

A2 Footprint A belongs to monster **3**.
Footprint B belongs to monster **1**.
Footprint C belongs to monster **2**.
Footprint D belongs to monster **1**.
Footprint E belongs to monster **3**.
Footprint F belongs to monster **2**.

B1

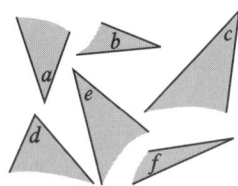

Angles *a*, *c* and *e* are equal to the unmarked angle.

B2 Angle *a* is equal to angle *f*.
Angle *b* is equal to angle *c*.
Angle *d* is equal to angle *e*.

C1

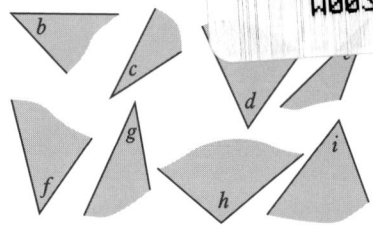

Bigger than *a*: *b*, *d*, *f*, *h*, *i*
Smaller than *a*: *c*, *e*, *g*

C2 *a* is the largest. *b* is the smallest.

D1 There are 8 right-angles in the diagram.

D2

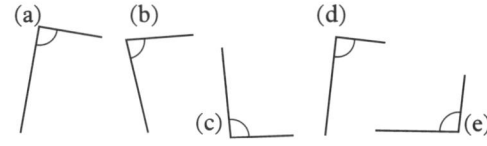

(a) and (d) are right-angles.

D3

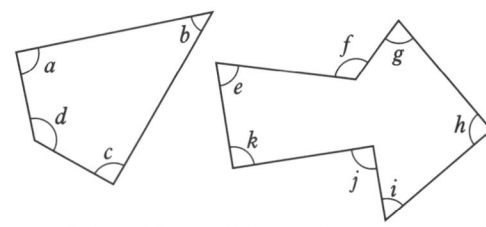

a, *c*, *h*, *j* and *k* are right-angles.

D4 (a)

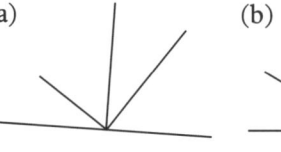

(b)

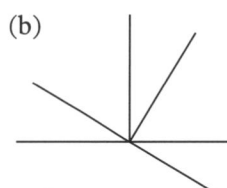

(a) There are 3 right-angles.
(b) There are 4 right-angles.

D5

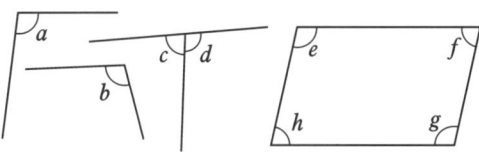

Bigger than a right-angle: *a*, *b*, *d*, *e*, *g*
Smaller than a right-angle: *c*, *f*, *h*

1

D6

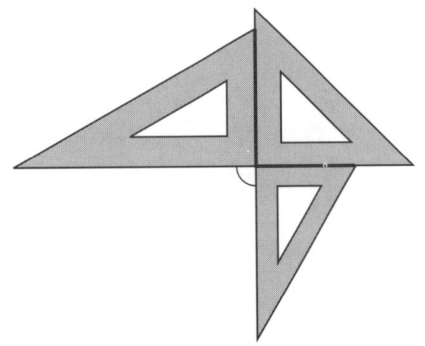

Yes, the marked angle is a right-angle.

D7

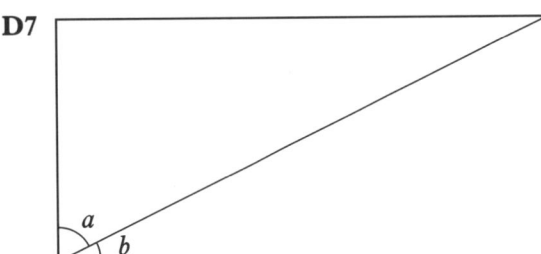

Angles a and b are not equal; a is bigger than b.

D8 Yes, angles c and d are equal.

D9

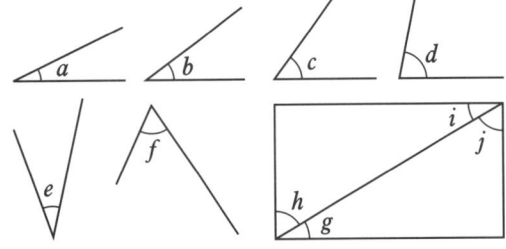

Bigger than $\frac{1}{2}$ of a right-angle: c, d, f, h, j
Smaller than $\frac{1}{2}$ of a right-angle: a, b, e, g, i

E1 There are 180 degrees in 2 right-angles.

E2 There are 270 degrees in 3 right-angles.

E3 There are 360 degrees in 4 right-angles.

In the next few questions your answers may be slightly different from the answers in this answer book. If an answer is only 1 or 2 degrees different it is probably a good answer.

E4 ▲ $a = 50$ degrees $b = 20$ degrees

E5 $a = 120$ degrees $b = 110$ degrees
      $c = 20$ degrees $d = 90$ degrees
      $e = 60$ degrees $f = 140$ degrees

E6 $a = 50$ degrees $b = 60$ degrees
      $c = 70$ degrees $d = 35$ degrees
      $e = 90$ degrees $f = 55$ degrees

E7 $a = 90$ degrees $b = 103$ degrees
      $c = 158$ degrees

E8 $d = 60$ degrees $e = 71$ degrees
      $f = 107$ degrees $g = 118$ degrees
      $h = 64$ degrees $i = 65$ degrees

Turning

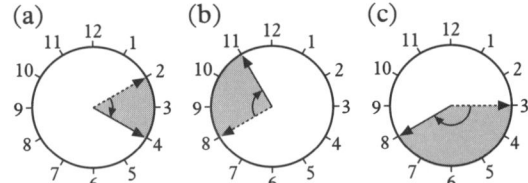

A1 ▲ The hour hand turns through 30 degrees between 7 and 8 o'clock.

A2 (a)   (b)   (c)

The hour hand turns through:
(a) 60 degrees   (b) 90 degrees
(c) 150 degrees

A3 Between 10 o'clock and 7 o'clock the hand turns through 270 degrees.

A4		Times	Angle turned through
	(a)	8 a.m. and 1 p.m.	150 degrees
	(b)	8 a.m. and 3 p.m.	210 degrees
	(c)	11 a.m. and 10 p.m.	330 degrees
A5	(a)	6 a.m. and 3 p.m.	270 degrees
	(b)	10 a.m. and 4 p.m.	180 degrees
	(c)	7 a.m. and 7 p.m.	360 degrees

A6 The beam turns through 22 or 23 degrees.

A7 The beam has turned through 35 or 36 degrees.

A8 (a) The beam will hit helicopter G first.
   (b) The beam will hit helicopter F second.
   (c) The beam will hit the other helicopters in the order:
        D, C, E, B, A.

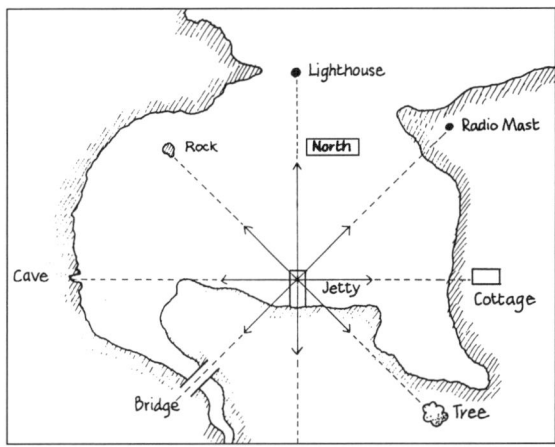

B1 The **cave** is due west of the jetty.

B2 The **bridge** is south-west of the jetty.

B3 The cottage is **due east** of the jetty.

B4 The man turns through **45 degrees**.

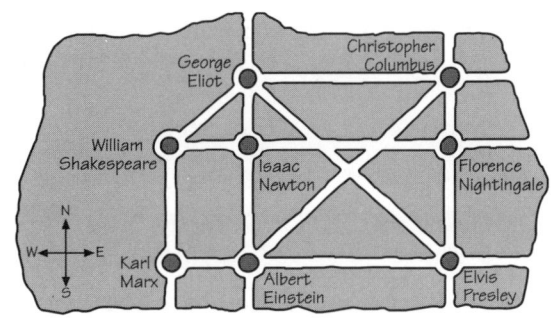

B5 ▲ These are the directions from Isaac Newton.
   (a) **George Eliot** is north.
   (b) **Florence Nightingale** is east.
   (c) **Albert Einstein** is south.
   (d) **William Shakespeare** is west.

B6 (a) If you stand by Karl Marx and face east you will see Albert Einstein.
   (b) You will also see Elvis Presley in the distance.

B7 You are facing south-west.

B8 (a) The girl is standing by George Eliot.
   (b) If she faces south-east she will see Elvis Presley.

B9 The picture shows Karl Marx.

B10 (a) Florence Nightingale is marked A.
   (b) The arrow points west.

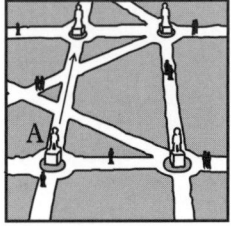

Answers 1 degree bigger or smaller than the answers given here should be marked correct.

C1 ▲ The bearing of London Airport from Southampton is **47 degrees**.

3

C2 ▲ These are the bearings from Maldon.

Colchester	45 degrees
Burnham	139 degrees
Southend	172 degrees
Rayleigh	193 degrees
Wickford	217 degrees
Chelmsford	270 degrees
Braintree	332 degrees

C3 (a) Bierton and Hulcott have a bearing of 50 degrees from Aylesbury.

(b), (c)

Place	Bearing from Aylesbury	Distance from Aylesbury
Bierton	50 degrees	2·2 km
Hulcott	50 degrees	4·4 km
Hardwick	344 degrees	5·1 km
Whitchurch	344 degrees	7·0 km
Pitchcott	326 degrees	8·0 km

D1 ▲

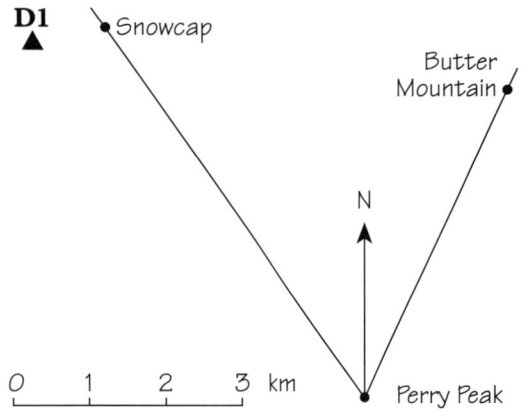

D2

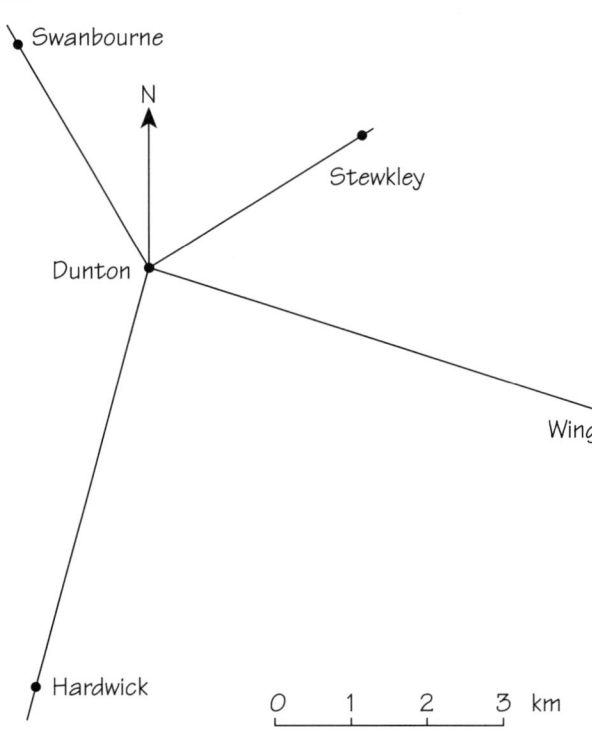

D3

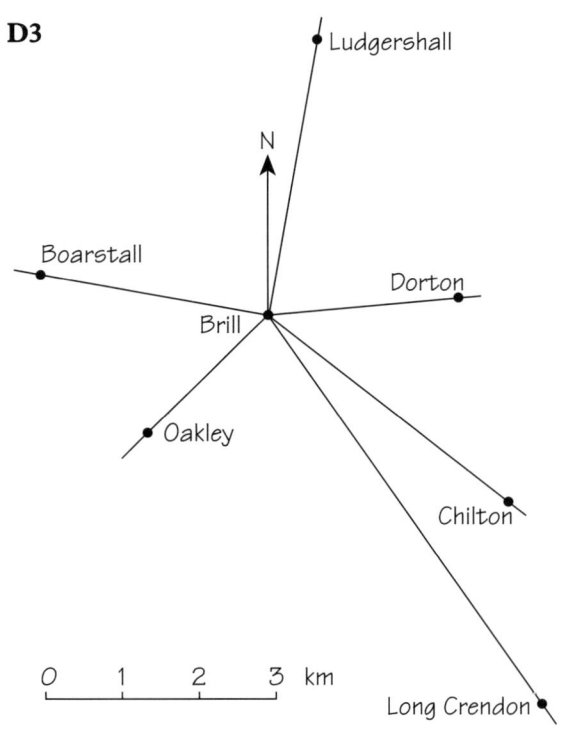

Area 1

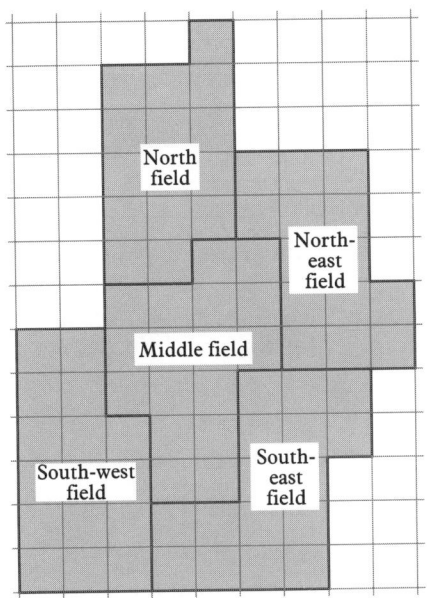

A1 and **A3**   The largest field is Middle field.

A2 and **A4**   North-east field is smallest.

A5   South-west and South-east fields are the same size.

A6   South-east field is marked A.

A7   Miss Wesby can keep 4 horses in South-east field.

A8   She can keep:
(a) 3 horses in North field,
(b) 4 horses in Middle field.

A9   If North and Middle fields are made into one field, she can keep 8 horses in the new big field.

A10

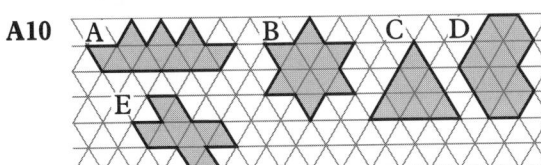

(a) The largest shape is D.
(b) D covers 14 triangles.
(c) Shape C is the smallest.
(d) A and B are the same size.

A11

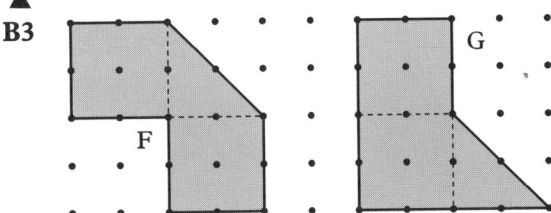

(a) D and E are the same size as A.
(b) C and F are the same size as B.
(c) A and E are congruent.

B1

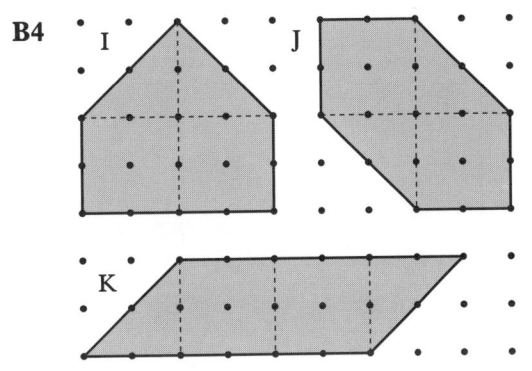

B2   C has the greater area.
▲

B3

B4

B5

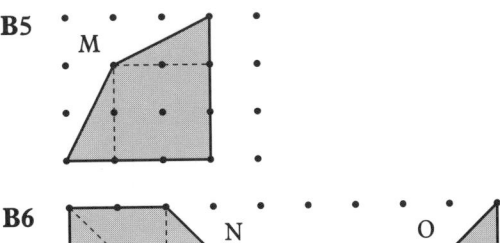

B6

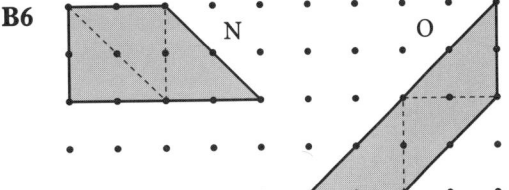

5

B7

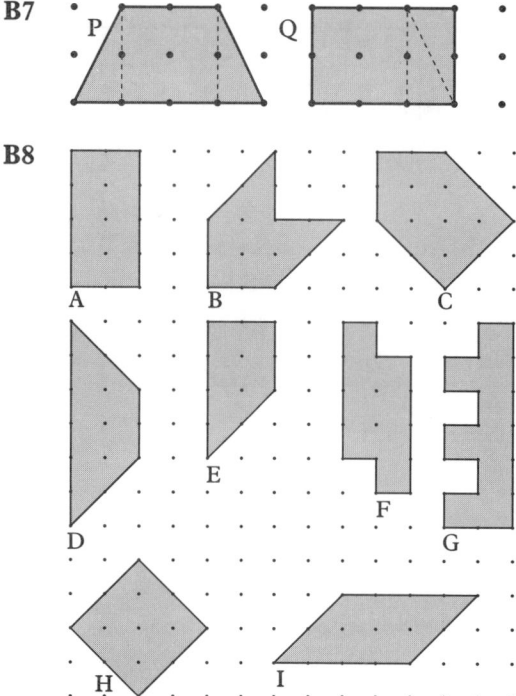

B8

B, D, F, H and I have the same area as A.

C1

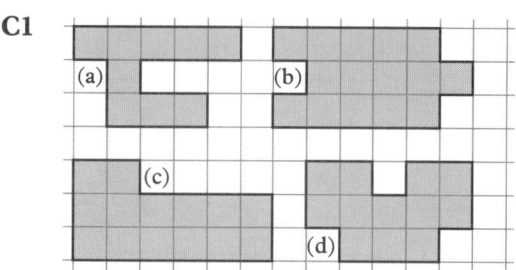

These are the areas of the shapes in your booklet.

(a) 9 sq cm   (b) 15 sq cm
(c) 14 sq cm   (d) 12 sq cm

C2 ▲
(a) 20 sq cm   (b) 17 sq cm
(c) 18 sq cm

C3 (a) 22 sq cm   (b) 17 sq cm

C4 (a) The area of the whole shape is 30 sq cm.
(b) The area of the coloured part is 15 sq cm.
(c) One-half of the whole area is coloured.

C5 (a) One-third of the area is coloured.
(b) Two-thirds is not coloured.

C6

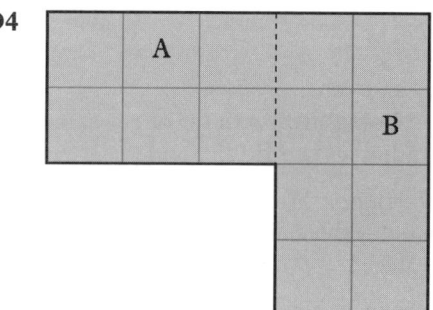

These are the areas of the shapes in your booklet.

(a) $5\frac{1}{2}$ sq cm   (b) 5 sq cm   (c) $6\frac{1}{2}$ sq cm
(d) 7 sq cm   (e) 8 sq cm   (f) 6 sq cm

C7 (a) $14\frac{1}{2}$ sq cm   (b) 11 sq cm
(c) 13 sq cm   (d) $14\frac{1}{2}$ sq cm
(e) 16 sq cm   (f) 15 sq cm

D1 The area of the rectangle is 15 sq cm.

D2 These are the areas of the rectangles.
(a) 16 sq cm   (b) 12 sq cm
(c) 12 sq cm   (d) 21 sq cm

D3 (a) The area is 10 sq cm.   (2 × 5 = 10)
(b) The area is 24 sq cm.   (4 × 6 = 24)
(c) The sides of the rectangle are 3 cm and 6 cm. Its area is 18 sq cm.
(d) The rectangle measures 4 cm by 7 cm. Its area is 28 sq cm.

D4

(a) The area of rectangle A is 6 sq cm.
(b) Rectangle B has an area of 8 sq cm.
(c) The area of the whole shape is 14 sq cm.

D5 The area of the whole shape is 26 sq cm.
Can you find two simple ways of dividing the shape to give the answer?

E1

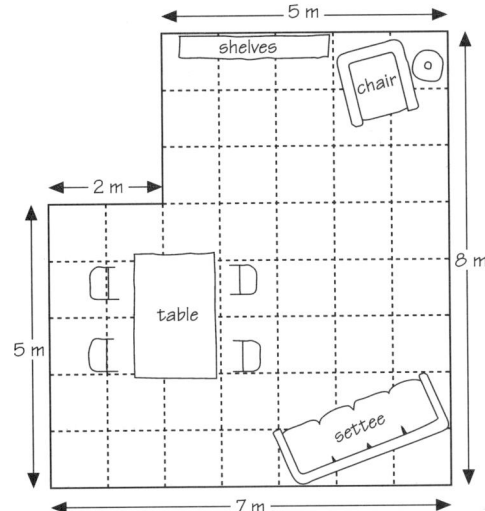

The area of the floor is 50 square metres.

Check your answer by dividing the floor up in a different way.

Scale drawing 1

A1 The short side of the piece of wood is 40 cm long.

A2 You must turn over one of the sides to make them fit.
Here is one way of doing it.
You may have done it slightly differently.

Scrap		
Top	Back	Side
Bottom	Front	Side

B1

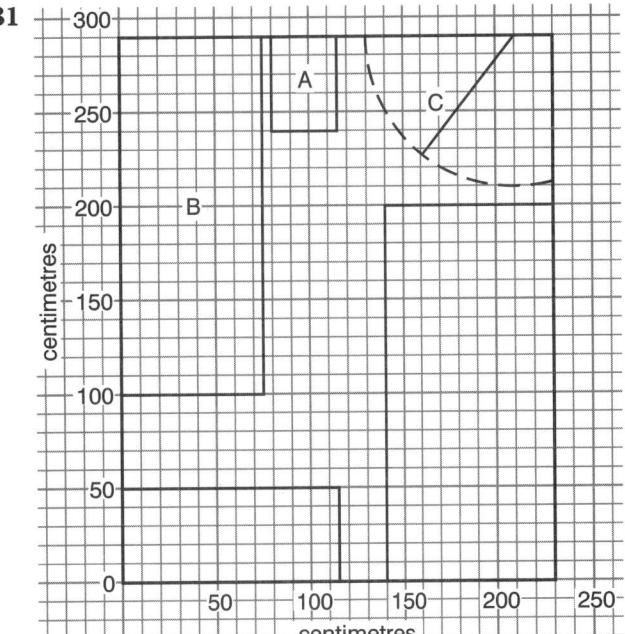

(a) Rectangle A stands for the bedside table.
(b) Dawn's bed is 75 cm wide.
(c) Dawn's bed is 190 cm long.
(d) The length of Kathy's bed is 200 cm.
(e) The gap between the beds is 65 cm.
(f) Line C stands for the door.
(g) The dotted line shows the path of the edge of the door.
(h) As you go into the room, Dawn's bed is on the right.

B2 (a) On the plan, the length of the room is 255 mm.
(b) The real room is 255 cm long.
(c) The real door is 80 cm wide.

B3

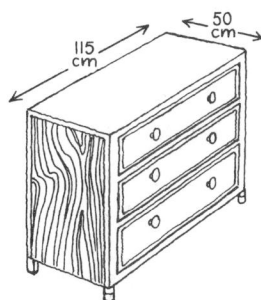

No, the chest will not fit along the wall between A and B. It is between 4 and 5 cm too long.

B4, B5 and **B6**

You can check that your plans are the right size by laying them on the drawings below.

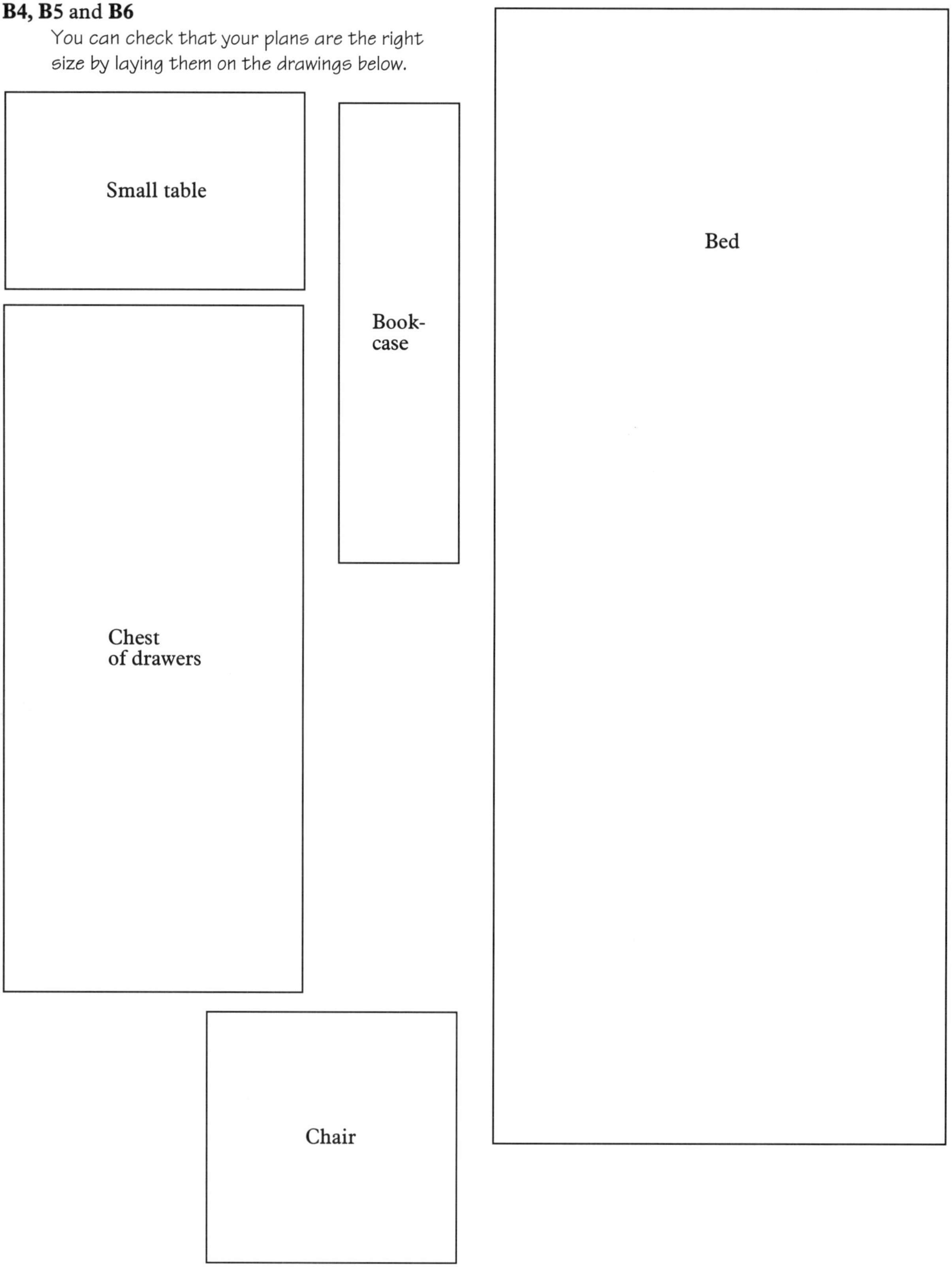

Small table

Book-
case

Bed

Chest
of drawers

Chair

B7 No, the bed is too long.

B8 (a) No, the chair and bookcase will not fit.
(b) Yes, the table, chair and bookcase will fit.

B9 (a) The bed gets in the way of the bookcase.
(b) The bed gets in the way of the drawers.

B10 Show your worksheet to your teacher.

Area 2

A1 These are the areas of the stamps.
(a) 9 sq cm   (b) 6 sq cm
(c) 12 sq cm   (d) 12 sq cm
(e) 24 sq cm

A2 These are the areas of the stamps. Their measurements are given in brackets.
(a) 4 sq cm   (2 cm by 2 cm)
(b) 9 sq cm   (3 cm by 3 cm)
(c) 15 sq cm   (3 cm by 5 cm)

A3 The areas of the tickets are:
(a) 21 sq cm   (b) 24 sq cm
(c) 45 sq cm

A4

$\frac{1}{2}$ of 7 is $3\frac{1}{2}$.

A5 $\frac{1}{2}$ of 9 is $4\frac{1}{2}$.

A6 $\frac{1}{2}$ of 3 = $1\frac{1}{2}$

A7

Number	1	2	3	4	5	6	7	8	9	10
$\frac{1}{2}$ of number	$\frac{1}{2}$	1	$1\frac{1}{2}$	2	$2\frac{1}{2}$	3	$3\frac{1}{2}$	4	$4\frac{1}{2}$	5

A8 $\frac{1}{2}$ of 25 = $12\frac{1}{2}$

A9 (a) $\frac{1}{2}$ of 15 = $7\frac{1}{2}$ (b) $\frac{1}{2}$ of 21 = $10\frac{1}{2}$
(c) $\frac{1}{2}$ of 19 = $9\frac{1}{2}$

B1 (a) The area of the rectangle is 12 sq cm.
(b) Half of the rectangle is shaded.
(c) The area of the shaded triangle is 6 sq cm.

B2

	Area of rectangle	Area of triangle
(a)	10 sq cm	5 sq cm
(b)	15 sq cm	$7\frac{1}{2}$ sq cm
(c)	3 sq cm	$1\frac{1}{2}$ sq cm
(d)	20 sq cm	10 sq cm
(e)	16 sq cm	8 sq cm
(f)	24 sq cm	12 sq cm
(g)	8 sq cm	4 sq cm
(h)	21 sq cm	$10\frac{1}{2}$ sq cm
(i)	12 sq cm	6 sq cm
(j)	14 sq cm	7 sq cm

B3 ▲

	Area of rectangle	Area of triangle
(a)	15 sq cm	$7\frac{1}{2}$ sq cm
(b)	6 sq cm	3 sq cm
(c)	8 sq cm	4 sq cm
(d)	5 sq cm	$2\frac{1}{2}$ sq cm
(e)	27 sq cm	$13\frac{1}{2}$ sq cm
(f)	25 sq cm	$12\frac{1}{2}$ sq cm
(g)	44 sq cm	22 sq cm
(h)	35 sq cm	$17\frac{1}{2}$ sq cm

B4

A   3 square feet   B $7\frac{1}{2}$ square feet
C   4 square feet   D   8 square feet
E   10 square feet   F   9 square feet
G   12 square feet

9

C1

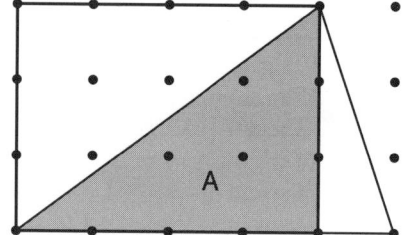

(a) The area of the rectangle is 12 sq cm.
(b) The area of triangle A is 6 sq cm.

C2

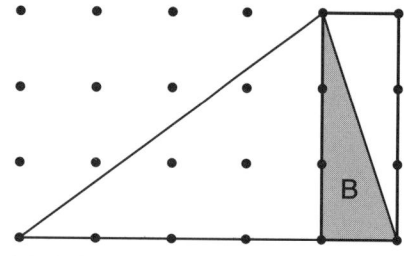

(a) The area of the rectangle is 3 sq cm.
(b) The area of triangle B is $1\frac{1}{2}$ sq cm.

C3 The area of the triangle is $7\frac{1}{2}$ sq cm.

C4 ▲ These are the areas of the triangles on worksheet 3–5.

(a) 10 sq cm   (b) 16 sq cm   (c) 9 sq cm
(d) 18 sq cm   (e) 10 sq cm   (f) 15 sq cm
(g) 9 sq cm   (h) 12 sq cm   (i) 25 sq cm

D1

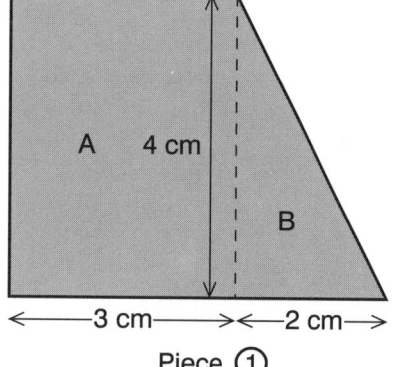

Piece ①

		Area
(a)	Rectangle A	12 sq cm
(b)	Triangle B	4 sq cm
(c)	Whole shape	16 sq cm

D2

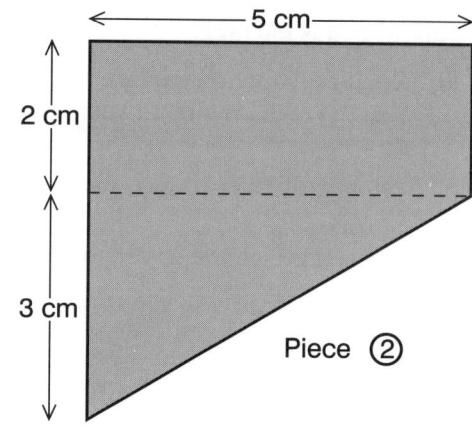

Piece ②

The area of piece ② is $17\frac{1}{2}$ sq cm.
(10 sq cm + $7\frac{1}{2}$ sq cm)

D3 The area of piece ③ is 69 sq cm.
(60 sq cm + 9 sq cm)

D4 ▲ (a) The area of the new design is $67\frac{1}{2}$ sq cm.

(b) It is smaller than the first design.

D5 These are the areas of the shapes on worksheet 3–6.

(a) 16 sq cm   (b) 20 sq cm
(c) 9 sq cm   (d) $13\frac{1}{2}$ sq cm
(e) 21 sq cm   (f) $27\frac{1}{2}$ sq cm
(g) $37\frac{1}{2}$ sq cm

If any of your answers are wrong, check your working carefully or try dividing the shape in another way.

Area 2: extension

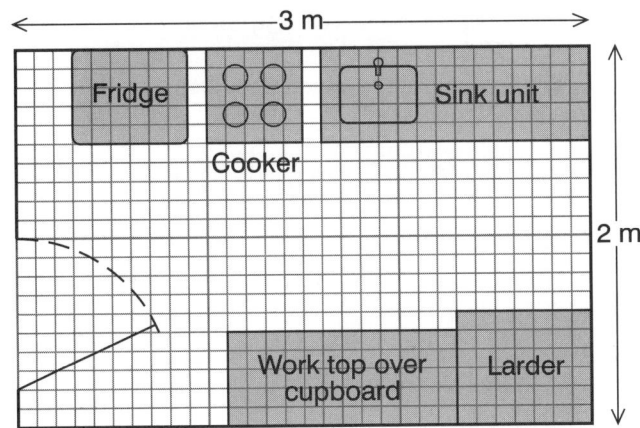

A1 The doorway is 0·8 metres wide.

A2 The width of the fridge is 0·6 metres.

A3 The length of the sink unit is 1·4 metres.

A4 The work top is 1·3 m long.

A5 (a) The cooker covers 25 small squares.
(b) The area of the cooker is 0·25 sq m.

A6

	Item	Area
(a)	Sink unit	0·70 sq m
(b)	Larder	0·42 sq m
(c)	Work top	0·65 sq m

A7 (a) The five things in the kitchen cover an area of 2·32 sq m.
(b) The area of the rest of the floor is 3·68 sq m.   (6 − 2·32 = 3·68)

A8 (a) $0.6 \times 0.4 = 0.24$ (b) $0.6 \times 0.8 = 0.48$
(c) $0.2 \times 0.4 = 0.08$ (d) $1.3 \times 0.2 = 0.26$
(e) $0.2 \times 0.3 = 0.06$ (f) $0.2 \times 0.2 = 0.04$

A9

	Size of rectangle	Area
(a)	0·8 m by 0·5 m	0·40 sq m
(b)	0·3 m by 0·3 m	0·09 sq m
(c)	0·7 m by 0·1 m	0·07 sq m

B1

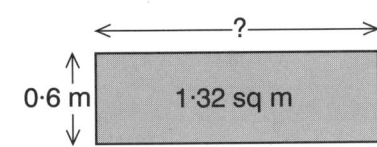

The other side is 2·2 m long.
(1·32 ÷ 0·6 = 2·2)

B2 The other side of the rectangle is 1·1 m long.   (1·54 ÷ 1·4 = 1·1)

B3

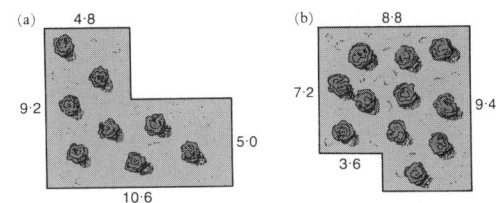

The perimeter of the pond is 8·6 m.

B4 The area of the pond is 4·32 sq m.

B5 and B6

(a) 4·8 9·2 5·0 10·6

(b) 8·8 7·2 9·4 3·6

Garden	Perimeter	Area
(a)	39·6 m	73·16 sq m
(b)	36·4 m	74·80 sq m

B7 (a) The other side of the rectangle is 1·7 cm long.
(26·8 − 23·4 = 3·4;   3·4 ÷ 2 = 1·7)
(b) The area of the rectangle is 19·89 sq cm.   (11·7 × 1·7 = 19·89)

B8 The perimeter is 7·4 m.
Hint: first find the length of the other side.

B9 The square has an area of 0·49 sq m.
Hint: first find the side of the square.

B10 These are the areas of the coloured shapes.
(a) 5·85 sq cm   (b) 7·29 sq cm
(c) 11·385 sq cm

B11 Challenge Problems

(a) The rectangle measures 4 m by 30 m.
You need to find by experiment two numbers whose product is 120 and whose sum is 34.

(b) This rectangle measures 6 m by 27 m.
How did you work your answer out?

C1

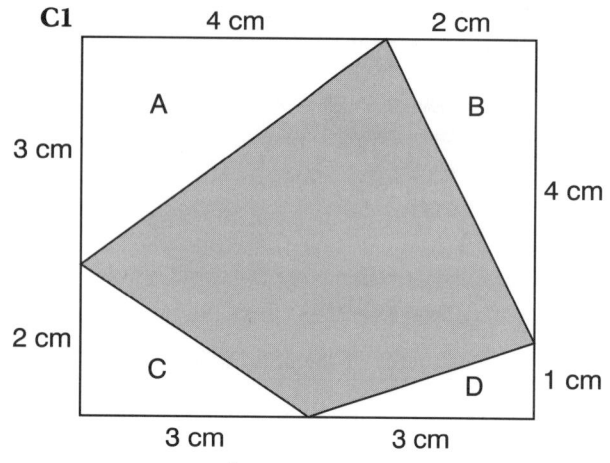

Triangle	Area in sq cm
A	6
B	4
C	3
D	1·5

C2 The area of the shaded quadrilateral in **C1** is 15·5 sq cm. (30 − 6 − 4 − 3 − 1·5 = 15·5)

C3 (a) The area of the shaded triangle is 11 sq cm.

       (b) The area of the shaded quadrilateral is 22 sq cm.

Remember to work out the areas of the rectangle and the right-angled triangles first.

C4

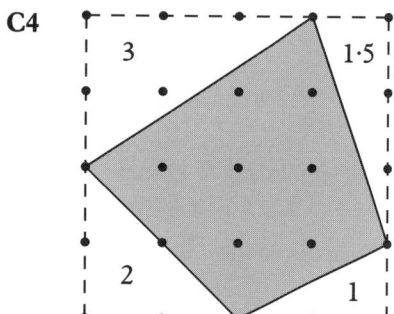

The area of the square is 16 sq cm.
The areas of the four triangles are 2 sq cm, 3 sq cm, 1·5 sq cm and 1 sq cm, a total of 7·5 sq cm.
So the area of the shaded quadrilateral is 16 sq cm − 7·5 sq cm = 8·5 sq cm.

C5 These are the areas of the shapes.
      (a) 8·5 sq cm   (b) 10 sq cm
      (c) 7 sq cm   (d) 9 sq cm
      (e) 8 sq cm
      Ask your teacher if you need help.

C6

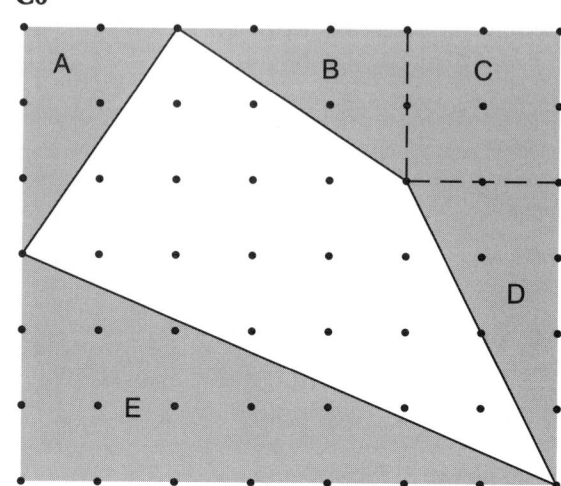

(a)

Shape	A	B	C	D	E
Area in sq cm	3	3	4	4	10·5

     (b) The area of the shape in the middle is 17·5 sq cm.
         (42 − 3 − 3 − 4 − 4 − 10·5 = 17·5)

C7 Here are the areas of the shapes.
      (a) 12·5 sq cm   (b) 13 sq cm
      (c) 16 sq cm   (d) 7·5 sq cm

D1 Here are the areas of the triangles.
      (a) 7·5 sq cm   ($\frac{1}{2}$ of 15)
      (b) 10 sq cm   ($\frac{1}{2}$ of 20)
      (c) 14 sq cm   ($\frac{1}{2}$ of 28)

D2 (a) 10·88 sq cm   (b) 7·84 sq cm
      (c) 24·295 sq cm

D3

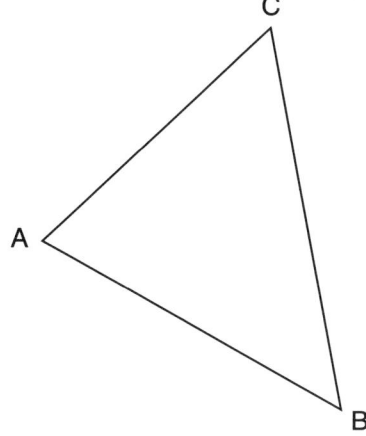

(a) AB = 4·5 cm
(b) Height = 4 cm
(c) Area of ABC = ½ of 4·5 × 4 sq cm
  = 9 sq cm

D4 and D5 *The area of the triangle does not change so you would expect your answers to be 9 sq cm each time. It is not possible to measure the lengths exactly. Answers between 8·6 sq cm and 9·4 sq cm are acceptable.*

D6

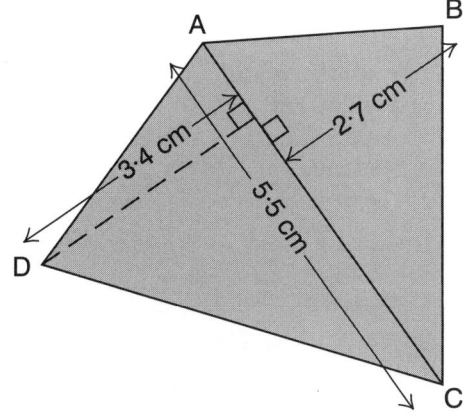

(a) Area of triangle ABC = 7·425 sq cm
(b) Area of triangle ACD = 9·35 sq cm
(c) Area of quadrilateral ABCD
  = 7·425 sq cm + 9·35 sq cm
  = 16·775 sq cm

D7 *Your answers to parts (c), (d) and (e) will depend on the measurements you get.*
(a) Length of PR = 5·5 to 5·6 cm
(b) Height of PQR = 2·3 to 2·5 cm
(c) Area of PQR = 6·3 to 7·0 sq cm
(d) Area of PRS = 11·2 to 11·9 sq cm
(e) Area of PQRS = 17·5 to 18·9 sq cm

D8 *Your answers to parts (d), (e) and (f) will depend on the measurements you get.*
(a) Length of BD on the plan is 6·9 to 7·0 cm.
(b) Real distance from B to D is 69 to 70 m.
(c) Height of triangle ABD (with base BD) is 42 to 43 m.
(d) Area of triangle ABD is 1440 to 1510 sq m.
(e) Area of triangle BCD is 2860 to 3000 sq m.
(f) Total area of field ABCD is 4300 to 4510 sq m.

D9 (a) Area of ABC, using AC as the base, is 1840 to 1990 sq m.
(b) Area of ACD, using AC as the base, is 2410 to 2580 sq m.
(c) Total area of the field is 4250 to 4570 sq m.

D10

Field	Area in sq m
North-west	4000 to 4300
South-west	3600 to 3950
South-east	7100 to 7600
North-east	7000 to 7500

E1 A good estimate for the area of the pond is 16 to 18 sq m.

E2

		Estimated area
(a)	Wallasea Island	9 to 11 sq km
(b)	Foulness Island	23 to 26 sq km
(c)	Potton Island	3·5 to 4·5 sq km
(d)	New England Island	1·5 to 2·5 sq km
(e)	Havengore Island	2 to 2·5 sq km
(f)	Rushley Island	0·7 to 0·9 sq km

Volume

A1

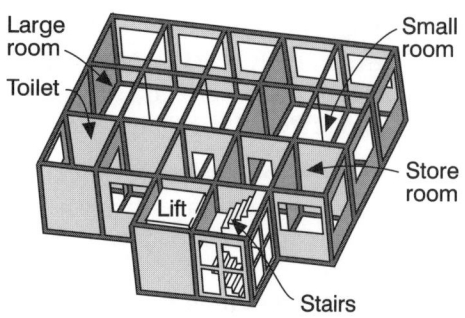

Large room, Toilet, Small room, Store room, Lift, Stairs

There are 17 cubes in this floor.

A2 If the building has 4 floors, there are 68 cubes in the whole building.
($17 \times 4 = 68$)

A3

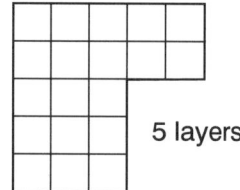

5 layers

(a) There are 19 cubes in one layer.
(b) There are 95 cubes in the whole building.

A4

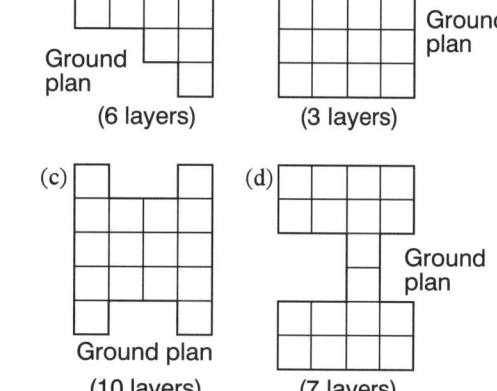

(a) Ground plan (6 layers)
(b) Ground plan (3 layers)
(c) Ground plan (10 layers)
(d) Ground plan (7 layers)

Here are the number of cubes in each building.
(a) 66 cubes   (b) 48 cubes
(c) 160 cubes   (d) 126 cubes

A5

(a) The area of the ground plan is 21 square cm.
(b) There are 63 centimetre cubes in the model.

A6

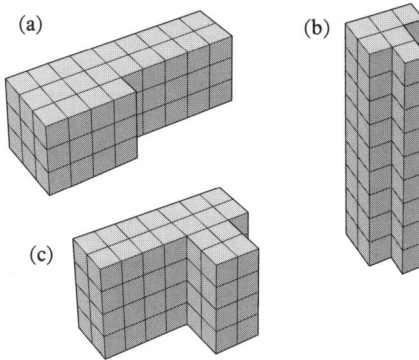

(a)   (b)   (c)

The number of centimetre cubes in each model is:
(a) 66   (b) 54   (c) 72

A7 Here are the volumes of the models.
(a) 6 cubic cm   (b) 12 cubic cm
(c) 12 cubic cm   (d) 80 cubic cm
(e) 48 cubic cm   (f) 27 cubic cm
(g) 38 cubic cm

A8

(a) The ground plan covers 26 squares.
(b) The volume of one layer is 26 cubic cm.
(c) The volume of the model is 104 cubic cm.
(d) The volume of the hole is 16 cubic cm.
(e) With the hole filled in, the volume would be 120 cubic cm.

A9 (a) Volume of hole   27 cubic cm
       (b) Volume of model   78 cubic cm
       (c) Volume with hole
           filled in   105 cubic cm

B1

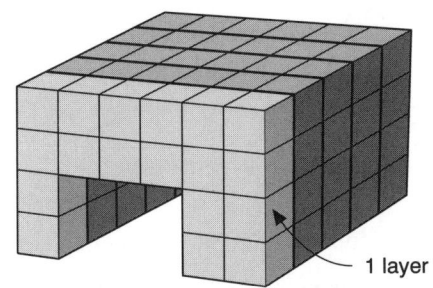

   ── 1 layer

(a) There are 18 cubes in one layer.
(b) The number of layers is 5.
(c) There are 90 cubes in the whole
    model. $(18 \times 5 = 90)$

B2 (a)   (b)

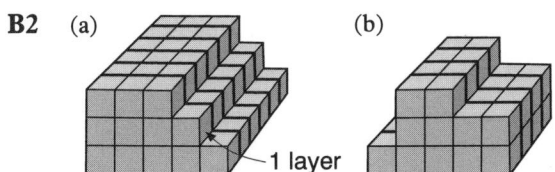

   ── 1 layer

The volumes of the models are:
(a) 72 cubic cm   (b) 44 cubic cm

	(a) Number of cubes in one layer	(b) Number of layers	(c) Volume in cubic cm
B3	10	5	50
B4	14	4	56
B5	16	5	80
B6	12	7	84
B7	15	3	45
B8	8	4	32
B9	14	8	112
B10	3	9	27
B11	16	3	48
B12	8	4	32
or	16	2	32

C1 (a) Yes   (b) No   (c) Yes
       (d) Yes   (e) No   (f) Yes
       (g) Yes   (h) Yes   (i) No
       (j) Yes

C2 (a)   (c)

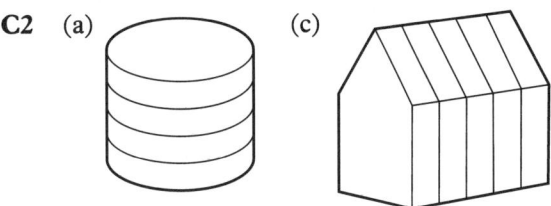

C3 (a) and (d) are prisms.

	(a) Area of cross-section	(b) Volume of each layer	(c) Volume of prism
C4	8 sq cm	8 cubic cm	24 cubic cm
C5	5 sq cm	5 cubic cm	30 cubic cm

D1 Here are the volumes of the prisms.
       (a) 56 cubic cm   (b) 21 cubic cm
       (c) 72 cubic cm   (d) 120 cubic cm
       Ask your teacher if you need help.

D2 (a) The volume of the prism is
           44 cubic cm.
       (b) The prism has a volume of
           45 cubic cm.

D3 (a) Volume of hole = 30 cubic cm
       (b) Volume with hole filled in
                   = 160 cubic cm
       (c) Volume of prism = 130 cubic cm

E1 A teacup holds about 40 medicine
       spoonfuls.

E2 You can fill 5 cups from a 1 litre carton.

E3 (a) 1 litre of water weighs 1000 grams.
       (b) 1000 grams is called a kilogram.

Scale drawing 2

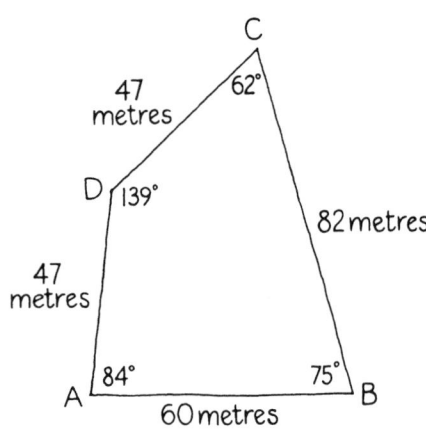

A1 The distance from A to C is between 87 and 89 m.

A2 It is between 71 and 73 m from B to D.

A3 You cannot measure from B to D inside the castle because Ye keep is in the way.

B1

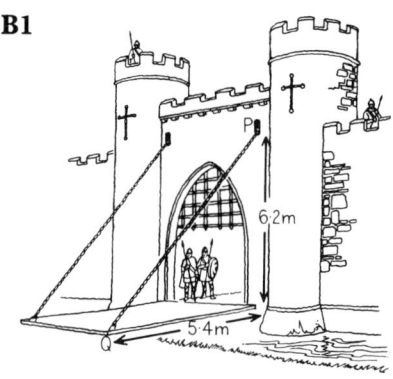

The chain PQ is between 8·1 and 8·3 m long.

B2 The angle between the chain and the ground is between 48° and 50°.

B3 When the drawbridge is raised through 40°, the new length of the chain is between 4·9 and 5·1 m.

B4 When the angle is 65°, the chain is between 2·5 and 2·7 m long.

B5 The dotted curve is a quarter of a circle.
▲

C1

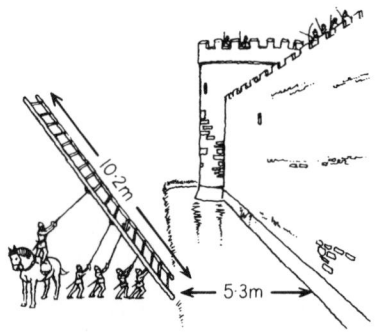

The ladder reaches 8·6 to 8·8 m up the wall.

C2 Where the moat is 6·7 m wide the ladder reaches 7·6 to 7·8 m up the wall.

C3 The longer ladder reaches 9·2 to 9·4 m up the wall.

C4 To reach the top of the wall the ladder must be 15·4 to 15·6 m long.

	Angle raised	Length of ladder
C5	70°	6·1 to 6·3 m
C6	85°	7·2 to 7·4 m

D1

The door is 6·4 to 6·6 m above the ground.

D2 The heights of the tops of the slopes are:
(a) 7·0 to 7·2 m (b) 2·9 to 3·1 m

D3 The angles of the slopes are:
(a) 23° to 25° (b) 33° to 35°

D4 The first escaper is 2·1 to 2·3 m below ground level.

D5 (a) The man is 3·2 to 3·4 m below ground level.
(b) The rabbit is 2·1 to 2·3 m below ground level.

E1 Show worksheet 4–1 to your teacher.
The plan shown here is a reduced copy of
the one on worksheet 4–1. The position of
each dagger is indicated by a cross.

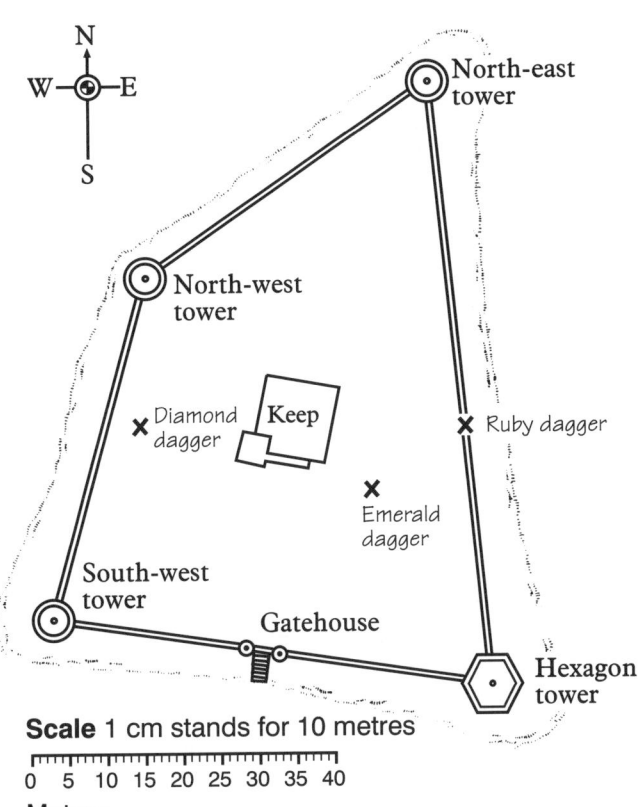

Scale 1 cm stands for 10 metres

0   5   10 15 20 25 30 35 40

Metres

Area 3

A1 George uses
26 cm of edge
strip.

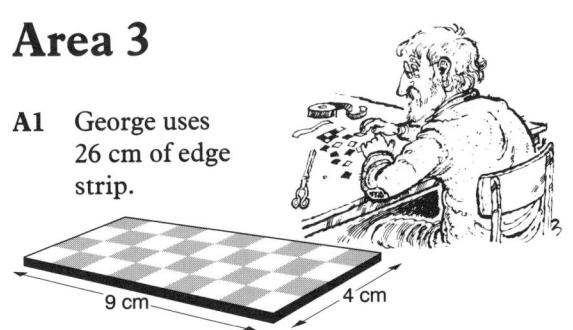

A2 (a) Your own diagram.
(b) Your own diagrams.
Your three rectangles should measure
1 cm by 36 cm, 2 cm by 18 cm, and 6 cm
by 6 cm.

A3

Size of rectangle	Number of of squares	Length of edge strip
9 cm by 4 cm	36	26 cm
12 cm by 3 cm	36	30 cm
6 cm by 6 cm	36	24 cm
18 cm by 2 cm	36	40 cm
36 cm by 1 cm	36	74 cm

A4 The 6 by 6 tray has the shortest edge strip.

A5

Size of rectangle	Number of of squares	Length of edge strip
1 cm by 24 cm	24	50 cm
2 cm by 12 cm	24	28 cm
3 cm by 8 cm	24	22 cm
4 cm by 6 cm	24	20 cm

A6
The distance all the way round the outside
edge of a shape is called the **perimeter**.

A7

	(a) Area in sq cm	(b) Perimeter in cm
A	18	18
B	18	24
C	18	22
D	18	38

A8 There are 12 possible diagrams. Here they
are, with their perimeters. Only one has a
perimeter of 10 cm.

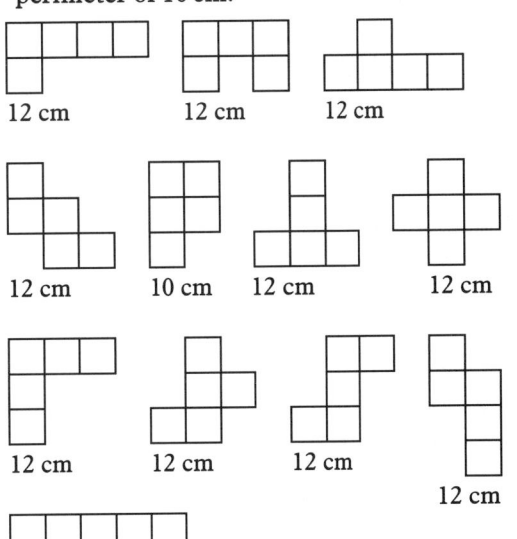

A9

Shapes with the same area may have different perimeters.

B1

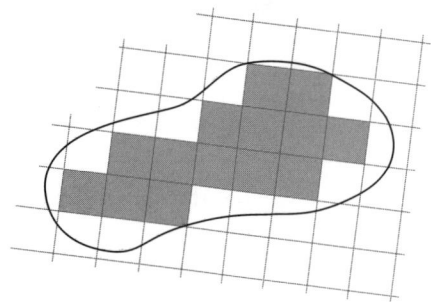

There are 14 whole squares inside the shape.

B2

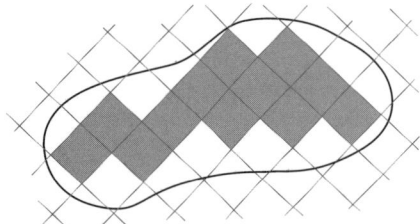

(a) There are only 12 whole squares now.
(b) The largest possible number of whole squares is 16.

B3 Here are the largest possible number of squares.
(a) 6   (b) 13   (c) 10

C1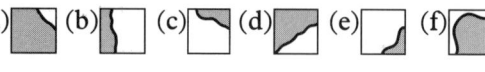

(a) 1   (b) 0   (c) 0   (d) 1   (e) 0   (f) 1

C2 (a) There are 8 pieces of squares.
(b) The rough area of the shape is 22 sq cm.

C3 (a) The area is about 12 sq cm.
(b) The area is about 21 sq cm.

C4 The shapes have area:
(a) 22 to 24 sq cm   (b) 17 to 19 sq cm

C5 (a)

	Area
Rumanian hamster	180 to 190 sq cm
Field vole	45 to 50 sq cm

(b) No, the hamster's skin is not twice the area of the vole's skin.
(c) *The hamster's skin is about 4 times the area of the vole's skin.*

D1

	(a) Number of sheep	(b) Number of hurdles
A	12	14
B	49	28
C	18	18
D	16	20
E	12	26

D2 (a), (b), (c)

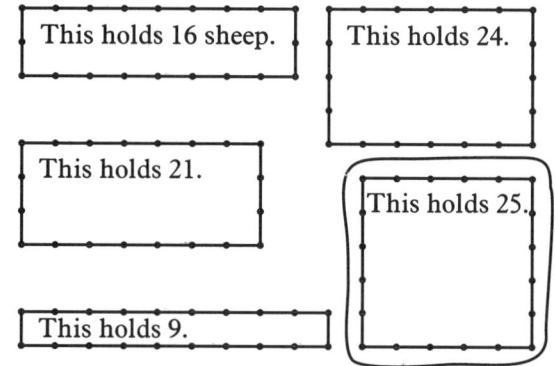

D3 (a) Cheryl can make 1 by 11,   2 by 10, 3 by 9,   4 by 8,   5 by 7   and 6 by 6 pens.
(b) The 6 by 6 pen holds most sheep (36 sheep).
(c) This pen holds 11 more sheep than the 5 by 5 pen.

D4 (a) The pen uses 58 hurdles.
(b) The smallest number of hurdles needed to pen 100 sheep is **40**.
What do you notice about the shape of this pen?

D5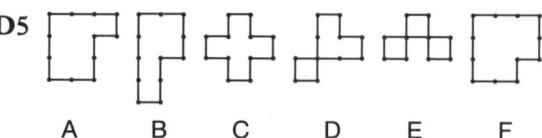

(a) **Pen F** holds the largest number of sheep.
(b) **Pen E** holds the smallest number.
(c) The list in order of size is:
     F,   A,   B,   C,   D,   E

D6 (a)

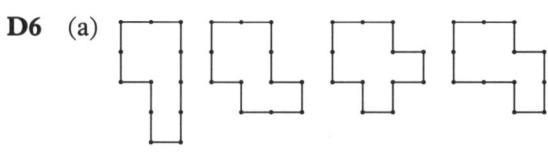

(b)

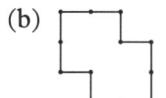

D7

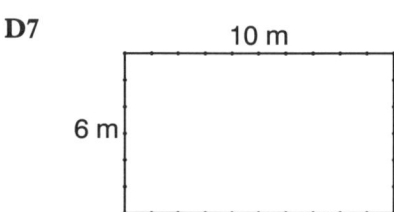

(a) Area = 60 sq m
(b) Perimeter = 32 m

D8 (a) Perimeter = 20 cm
Area = 28 or 29 sq cm
(b) Perimeter = 20 cm
Area = 22 to 25 sq cm
(c) Perimeter = 20 cm
Area = 16 to 20 sq cm

D9

> Shapes with the same perimeter may have different areas.

E1 (a)–(e) The perimeter of all five shapes is 24 cm.

E2

	Number of sides	Perimeter in cm	Area in sq cm
(a)	3	24	27 to 29
(b)	4	24	36
(c)	6	24	41 to 42
(d)	8	24	43 to 44
(e)	12	24	44 to 46

If you get different answers from these, discuss them with your teacher.

E3 The perimeter remains the same. As the number of sides increases the area increases.

F1 Karl used the wrong units. The area should be in sq cm and the perimeter in cm.

F2 (a) Area = 16 sq m   Perimeter = 16 m
▲ (b) Area = 24 sq m   Perimeter = 24 m
(c) Area = 28 sq m   Perimeter = 23 m
(d) Area = 52 sq m   Perimeter = 34 m

G1 (a) The total area of the white and grey pieces is 104 sq cm.
(b) The white pieces have an area of $49\frac{1}{2}$ sq cm.
(c) The total area of the mosaic is about 140 sq cm.   (14 × 10)
So the area of the black parts is about 36 sq cm.   (140 − 104)

Pythagoras' rule

A1

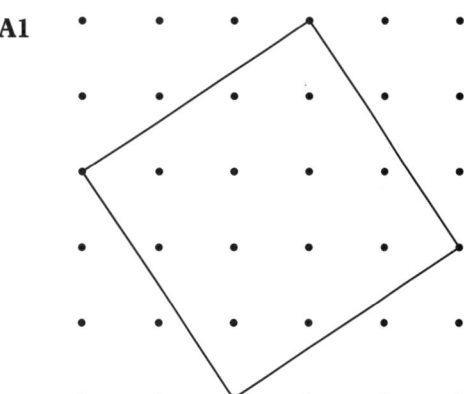

The area of the square is 13 sq cm.

A2 Here are the areas of the squares on worksheet 4–9.
(a) 10 sq cm   (b) 20 sq cm
(c) 8 sq cm   (d) 17 sq cm
(e) 29 sq cm   (f) 26 sq cm
(g) 25 sq cm

B1 The value of $\sqrt{10}$ is 3·16, so the square's side should be about 3·1 or 3·2 cm long.

B2 Here are the figures from the square root table. Your measurements in centimetres should be close to these.
(b) 4·47   (c) 2·83
(d) 4·12   (e) 5·39
(f) 5·10   (g) 5

B3 The side of the square is 5·48 cm (to 2 decimal places).

C1

Diagram	Area of P	Area of Q	Area of R
(a)	9 sq cm	4 sq cm	13 sq cm
(b)	4 sq cm	4 sq cm	8 sq cm
(c)	16 sq cm	9 sq cm	25 sq cm
(d)	1 sq cm	4 sq cm	5 sq cm
(e)	20 sq cm	5 sq cm	25 sq cm
(f)	1 sq cm	9 sq cm	10 sq cm

C2 (a)

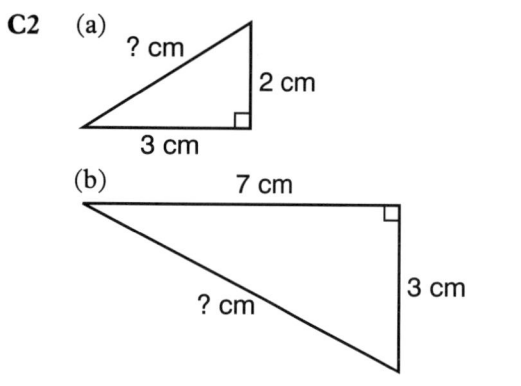

(a) 3·61 cm (√13) (b) 7·62 cm (√58)

C3 (a) 7·28 cm (√53) (b) 8·54 cm (√73)

C4 The length of the unknown side is
√20 cm = 4·47 cm.

C5 (a) 4 cm (b) 4·12 cm (c) 9·22 cm
(d) 6·24 cm

C6 (a) 5·39 cm (b) 7·48 cm (c) 9·54 cm
(d) 9·80 cm (e) 8·94 cm

C7

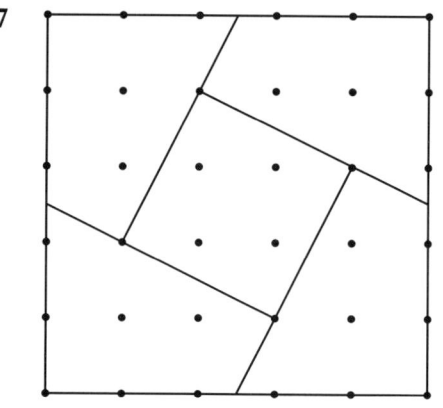

Measures

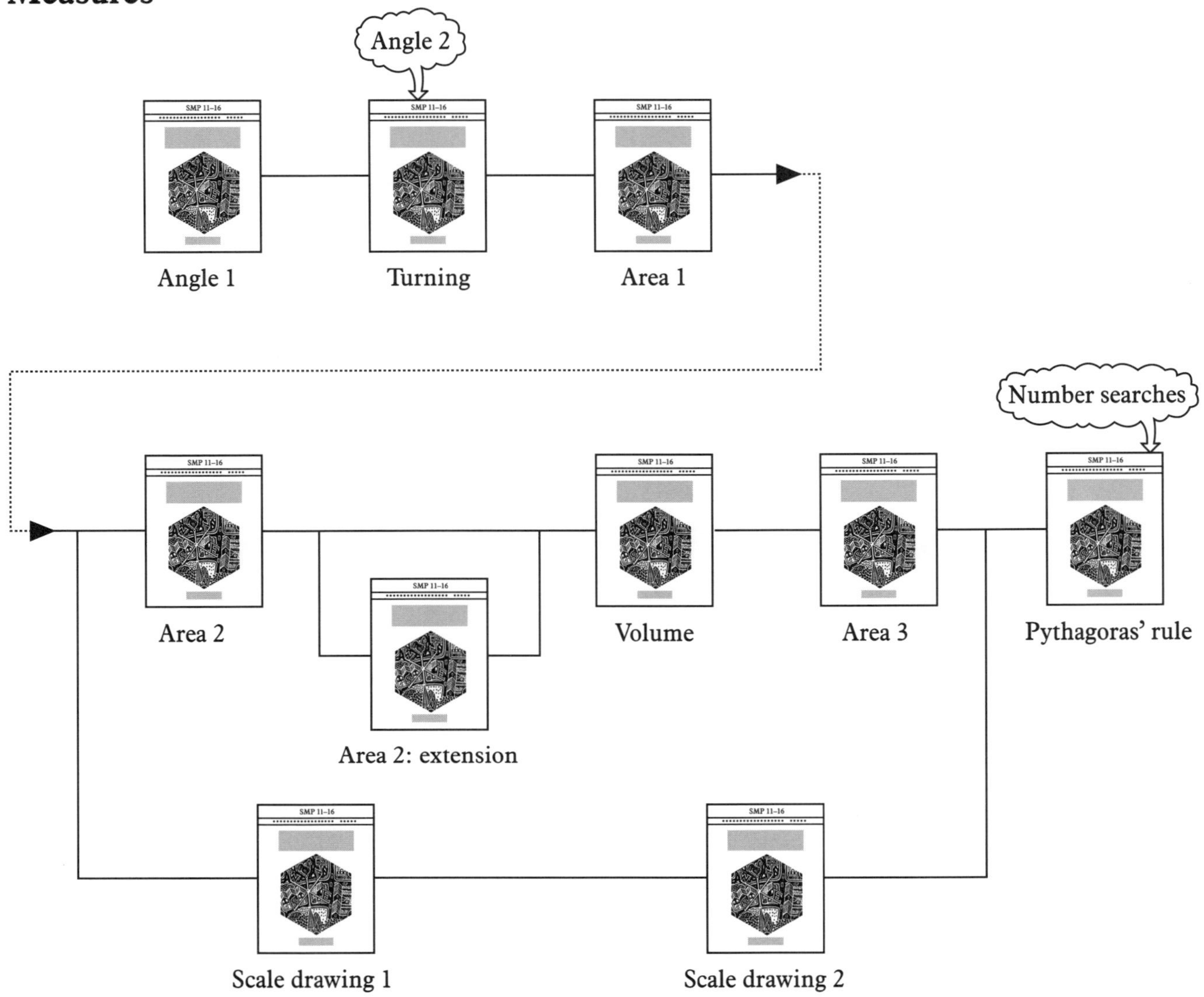

Angle 1 — Turning — Area 1

Angle 2 (→ Turning)

Area 2 — Area 2: extension — Volume — Area 3 — Pythagoras' rule

Number searches (→ Pythagoras' rule)

Scale drawing 1 — Scale drawing 2

SMP 11-16